Management Division 2018

Core Programming Area at the 2018 AIChE Spring Meeting and
14th Global Congress on Process Safety

Orlando, Florida, USA
22 – 25 April 2018

ISBN: 978-1-5108-6441-2

Printed from e-media with permission by:

Curran Associates, Inc.
57 Morehouse Lane
Red Hook, NY 12571

Printed by Curran Associates, Inc. (2018)

For permission requests, please contact AIChE
at the address below.

AIChE
120 Wall Street, FL 23
New York, NY 10005-4020

Phone: (800) 242-4363
Fax: (203) 775-5177

www.aiche.org

Additional copies of this publication are available from:

Curran Associates, Inc.
57 Morehouse Lane
Red Hook, NY 12571 USA
Phone: 845-758-0400
Fax: 845-758-2633
Email: curran@proceedings.com
Web: www.proceedings.com

TABLE OF CONTENTS

Some abstracts in this book refer to images that were not made available by the publisher

519831 "You were born with Two Ears and One Mouth for a Reason...": The Art of Listening for Understanding *(A GROW University Workshop)*

Abstract:

"You were born with Two Ears and One Mouth for a Reason ...": The Art of Listening for Understanding *(A GROW University Workshop)*

We spend about 70-80% of our waking hours in some sort of communication, 45% of which is listening! However, we are often listening with the intent to break into the conversation with our reply rather than to first truly understand. This workshop was originally part of a grassroots professional development series at Dow (called GROW U) that took the concepts found in Stephen Covey's Seven Habits of Highly Effective People, but looked at them in fresh ways. The workshop is centered on the concept of empathic listening from the 5th habit and will help you discover how to be an effective listener at work, leading to understanding of the real problems that need to be addressed in order to drive towards effective solutions.

The session will also provide an overview of how a group of Dow women and interested men (with management support) initiated a monthly workshop series called ***GROW University*** focused on personal and professional development. The topics are chosen by the group and the content of each workshop is developed and presented by small teams. This grassroots training program elevates the entire organization through the highly interactive workshops (which are open to everyone), while providing opportunity for individuals to demonstrate leadership, teaching ability, and passion for their chosen topics. The ultimate goal is to engender a higher level of satisfaction and productivity within a diverse and inclusive community. This concept was originally launched at one Dow site and has now gone global, indicating the inherent need for these types of growth opportunities.

512543 Role of Qualitative Risk Assessments in Managing Project Start-Ups

Abstract:

A qualitative approach to performing risk assessments for project start-ups has recently been developed at Oak Ridge National Laboratory with the goal of providing a simple tool to help manage projects that require cross-functional teams to implement a project with a variety of drivers and/or constraints. This approach focuses on the development of risk-based planning to guide selection of tasks to be performed during the planning and start-up phases of a project to reduce risk, lower costs, and improve chances of success. Examples of how the tool has been successfully implemented will be given.

511697 Commissioning Planning - the Key to a Successful Project Startup

Abstract:

In most industrial capital investment projects, the final stage of manufacturing technology delivery involves some combination of factory acceptance test, site acceptance test and post installation support. The intent of these services is to allow buyers of manufacturing systems to assure that systems meet specifications and lack any performance defects. Performance, as defined in nearly all specifications, is based upon nominal conditions. Unfortunately, variation thrives in factories. The combination of raw material variance, human error, environmental factors, infrastructure fluctuations and process interdependencies can cause significant performance issues within newly installed manufacturing systems. Successful projects account for these natural fluctuations in processes and establish more than equipment acceptance tests, they incorporate a commissioning plan. The commissioning plan recognizes the importance of running new systems in their natural environment, complete with variation. Deliberate testing beyond process boundary conditions allow startup teams to establish a set of process controls that ensure high quality production at required throughput rates. This presentation will offer a template for developing successful commissioning plans and provide a few examples of industrial operations that leveraged effective commissioning.

513362 3 Superpowers to Boost Your Career

Abstract:

Career growth has little to do with how smart you are or how hard you work. Despite the long hours, we stay in our role, wondering when we will get promoted or when our efforts will be recognized. Often we believe that others are holding us back, preventing us from getting that promotion or recognition and causing us stress. The truth is that, beyond a performance threshold, success has more to do with ourselves holding us back. The good news about this is that we can control ourselves. In this workshop, I will share three key concepts and techniques that I have learned that have helped me move from a worker to a leader - completely changing my career trajectory and limitations as a result.

Engineers are often hired for their knowledge or technical skills. But over the course of a career, they need to develop professionally - developing new skills and changing others' perception in order to gain visibility and to be given the opportunities to expand professionally. The path to this career success can be distilled down to 3 core behaviors, which I call superpowers. They are not obvious on the surface, but they make a huge difference when put into practice. First, change your relationship with fear. By changing our relationship with fear we change the decisions that we make. Second, change your story. This changes the perception of your ambitions and contributions. Third, learn to breathe. The ability to take a breath and then act in a considered way helps you deal with the challenges of change and it also builds others' confidence in you.

Since beginning to employ these strategies, my career has taken off. I made significant contributions outside of my role. Over the next few years I was granted an $85,000 scholarship from my company to pay for an Executive MBA, at age 30 I was named a Leader Under 40 by Control Engineering magazine, and I was promoted 3 times in 4 years up to a Global Director role at age 33.

514041 Ethics As the Underpinning of Safety and Process Safety

Abstract:

Ms. Grubbe will discuss the various relationships between personal, professional and business ethics and the implementation of successful occupational and process safety efforts. She will pull data from known disasters, as well as examine the root causes of incidents and outline some "ethical failures." This examination will lead into the link between the corporate "tone at the top" and the underlying mindset of the executive team. As the leader of the AIChE Center for Ethical Practice and as a retired corporate executive, the speaker will be speaking from her four decades of experience.

514014 Multiple Ways to Failure - How Harassment Hurts the Engineering Profession

Abstract:

Deborah Grubbe will address the subject of harassment in the engineering workplace from the following points of view: 1) the corporate executive (male), 2) the male engineering or project manager, 3) the woman engineering executive/manager, 4) the woman engineer, 5) the woman engineer harassing the male engineer, and 6) the male engineer. She will address questions from her 40 years of experience as an engineer, a manufacturing and engineering executive, an HR manager, and a sponsor and participant of identity networks of all types. Deborah has served on the Corporate Partnership Council for the Society of Women Engineers and as a Board Member of the Women in Engineering ProActive Network. She has held a number of multinational executive roles and is currently a member of the National Academy of Construction. She is the current leader of the AIChE Center for Ethical Practice.

515420 Remarks on Ethics in Patenting

Abstract:

A patent applicant has a duty of candor in filing papers with the U.S. Patent and Trademark Office. This duty can implicate, for example, the need to be honest when presenting data, and particularly not falling into the trap of excluding less-favorable data when trying to demonstrate unexpectedly good results to the USPTO. There are some notable cases in which failure to be honest with the USPTO, such as by failing to present all the data, led to a patent being entirely invalidated as unenforceable.

518350 Hiring People with Disabilities: Competitive Advantage for the Savvy Manager

Abstract:

John Johnston (PhD, MBA) is the past-chair of the American Chemical Society's Chemists with Disabilities Committee.

John developed the "Scientists with Disabilities" poster series to illustrate the vast potential of persons with disabilities.

These posters have been displayed in numerous government agencies and schools throughout the U.S.

As a means to manage his speech disability, John studied public speaking for many years and has become an award-winning speaker.

John will share an abridged version of his thought-provoking presentation "Hiring People with Disabilities: Competitive Advantage for the Savvy Manager"

with AIChE at the Spring 2018 national meeting.

518356 Navigating the ADA

Abstract:

Sandy is a licensed attorney in both Delaware and Pennsylvania. Throughout her professional career she has had the opportunity to focus on civil rights, and more particularly the ADA people with disabilities. When she first began her career it was a learning experience, and also a great opportunity to use her knowledge and training to make a difference in employment rights for a very under-represented segment of society. Her career began with working as a staff attorney at Delaware Community Legal Aid for the Disabilities Law Program. She knew she had a calling in the area of civil rights after working in the Volunteer Legal Services clinic while in law school.

With the passage of the ADA inn 1990 Sandy began further research, study and direct work for the rights of people with disabilities. As the years have passed, she has also focused on unconscious bias in all areas of civil rights for protected classes. A message she'd like to convey to those she has the pleasure of speaking with is that people with disabilities still have the highest unemployment rate, are entitled to reasonable accommodations in employment, and are subject to the highest rate of bias. We need to move the needle forward and take advantage of the vast talent of people with disabilities. Her focus for her discussion will be on those with disabilities who are capable of bringing much talent, productivity and advancement to employers in this field. The law provides for this. Navigating the ADA can be difficult at times.

She will focus on who is covered by the ADA, all aspects of the employment process, and particularly how to navigate to reasonable accommodation provisions of the ADA.

Within the vast population of people with disabilities, there are individuals with both visible and invisible disabilities. Those with invisible disabilities have formed some of their own support group networks as they seek to obtain gainful employment.

When Sandy was at what would have likely been critical years in her career she was diagnosed with a serious auto-immune disease. So at this point she was advocating for herself and it became more personal. It remains personal based on the relationships she has formed throughout her career.

She transitioned from tradition courtroom law to working in Civil Rights program management and leadership roles in Delaware State Government. Her current title is Statewide Administrator for Equal Employment Opportunity. She has had to opportunity to focus on individuals with disabilities. She was also part of the former Governor of Delaware, Jack Markell's state award winning platform of "Doing the Right Thing: Employment of People with Disabilities". This was Governor Markell's platform when he chaired the National Governor's Association during his tenure.

She continues to be the statewide lead on the American's with Disabilities Act. She regularly works with individuals with disabilities who are seeking employment opportunities. She also spends a tremendous amount of time working on accommodations for employees with disabilities.

Hiring individuals with disabilities is not just the right thing to do. It is also a source of untapped incredible talent from dedicated individuals who want to see their workplace succeed and to contribute to society with their talents and dedication.

During her presentation she will focus on The ADA requirements and the realities that individual's with disabilities often encounter. She believes that the reasonable accommodations guaranteed through the ADA should be discussed and further explained.

513484 Young Managers with Experienced Employees

Abstract:

There may be times throughout the career of a young engineer in which they will be placed in a role in which their colleagues - who have more experience - are reporting to them. This session will touch on the issues that can arise from this situation and what we can do (as young managers) to steer and control the relationship in order to maximize productivity.

Brandon Harding is currently a Chemical Process Engineer at the Holston Army Ammunition Plant (operated by BAE Systems) located in Kingsport, TN. In just 2 short years with BAE Systems, Brandon has been able to successfully implement several projects that save millions of dollars annually and has been awarded multiple Business Leader & Chairman Awards. As a process engineer, there are times when you do not directly manage colleagues - particularly those with more experience - but will require them to work for you in order to drive change. Brandon will discuss some personal experiences in which he had do "manage" those with more experience in order to accomplish a task or complete a project. Specifically, we will discuss how each issue began and what was done to control the situation and direct all project members toward a more productive environment. This will be an interactive portion (with the audience) of the session.

510755 The Sartomer Success System (3S): Achieving Operational Excellence

Abstract:

The Sartomer Success System (3S) was initiated at the Sartomer Chatham facility in June of 2015 and, based on that success, deployed at the Sartomer West Chester facility in June, 2017. The 3S Process is a site-wide Operating Excellence Initiative, creating a culture that allows our most valuable asset, our employees, to directly contribute to the success of the site. The 3S Process is used to identify and create employee-driven Pillar Teams, each with specific elements to address. These Pillar Teams are site specific, identified by the Plant Leadership Team, to address specific needs of the site aligned with the Company and Site Vision. For West Chester, these teams include: Process Safety, Daily Management Systems, 5S, Training, Focused Improvement, and Planned Maintenance, in addition to the 3S Steering Team. The 3S Steering Team is comprised of each of the Pillar Team Leaders, Plant Management Staff, and the Plant Manager. To date, more than 90% of our employees at Chatham (75% at West Chester) have voluntarily participated on one or more Pillar Teams.

Prior to 3S Deployment, Chatham struggled to engage hourly employees in improving the site's performance. Morale at the site was poor and product quality suffered. Chatham had experienced a number or process safety incidents, with poor self-reporting of smaller incidents (including near miss incidents), and had systemic issues with a number of other process safety (PSM) elements. Communication, which includes communication between management and the hourly associates, was also cited as an opportunity in employee surveys. For West Chester, similar gaps existed, with a greater need to improve onboarding of new employees due to a large number of employee retirements.

Since deploying 3S, Chatham has achieved "best ever" levels of personnel and process safety, product quality, cost control, and innovation. Similar gains are expected for West Chester. From a safety perspective, Chatham recently met the milestone of Sartomer's "double zero", which is more than 1 year without a major process safety incident (loss of primary containment or environmental release) or an OSHA recordable injury. Chatham has not had an OSHA recordable in more than 4 years, and has operated more than 11 years since it's last lost time incident. West Chester also reached more than 1,000,000 man hours since its last lost time injury earlier this year and is preparing for certification by OSHA's voluntary protection program (VPP) in early 2018. A business objective for Sartomer to achieve a "double zero" across the global business (4 manufacturing sites) in 2018.

The 3S Process is being used as a benchmark for Arkema, with many of the 3S elements being incorporated in to Arkema's own global Operational Excellence Initiative, SMART. Chatham was selected as a pilot facility for SMART in the Americas in early 2016, with the Plant Manager of Chatham joining the SMART design team mid-year, participating in the EU introduction of SMART at the EU Plant Manager's meeting, and assisting and creating the program for the SMART introduction at the America's Plant Manager meeting in late October, 2016. West Chester is deploying SMART concurrently with 3S starting in October, 2017.

Each Pillar Team's progress is displayed prominently on Pillar Boards, which includes: team membership (with pictures), the team's mission statement and objectives, key performance indicators (KPI's) to track monthly progress against these objectives, minutes, action items, and path forward. The Pillar Boards are maintained by the respective Pillar Team. This is to ensure visibility and accountability for all Pillar Teams, as well as encouraging further employee participation.

VSP, a Sartomer manufacturing site in France, also volunteered to be a SMART Pilot facility at the EU Plant Manager's meeting. The 3S Process is also planned for expansion across Sartomer at the Nansha (China) facilities. In short, all Sartomer facilities are currently building on processes learned with 3S, elements of which are being expanded across Arkema with the global deployment of SMART.

514228 1st Year Experience of a Young Manager in a Chemical Manufacturing Unit

Abstract:

Donna Bryant has been Unit Superintendent of the Environmental Operations Unit at Syngenta in St. Gabriel, LA for a little over a year. In this time she has learned a lot about managing people, especially people that are older than her. In this presentation, she will be discussing what she has learned over this past year, what she wishes she had known when she started, and what skills she plans to expand upon.

512479 An Engineer's Guide to Marketing Yourself Professionally

Abstract:

Everyone knows engineers are great at science, facts and figures, building products, and discussing technical concepts, but when it comes to "fluff" like communicating their story and building a personal brand they are out of their comfort zone. Regardless of industry or job function, personal branding both digitally and in the office has become an essential element of career success for building your networking, moving up the pipeline, and getting noticed for you achievements. In this session we will take the "art of personal branding" and make it a science with a practical guide of tips, tools, processes and formulas engineers can use to best build their professional brand, craft their story, and communicate their value in order to be recognized for their work accomplishments, grow their network, gain a promotion, or make a new career transition.

512322 Leadership By Engineers and Scientists

Abstract:

Nearly all engineers and scientists work in teams where a leader oversees and guides process/product development and direction. When technically-trained individuals undertake a leadership role, frustration is a frequent outcome, despite technical competency and good intentions. This talk will discuss reasons why engineers and scientists often find adaptation into leadership roles disconcerting and will explore ways to smooth the transition.

AUTHOR INDEX